BREATH OF LIFE

Growing Closer to God &
Growing a Secure Community
by Loving Creation

REV DELE

Soil & Souls Press
Marlboro, New York

Breath of Life: Growing Closer to God, Growing a Secure Community by Loving Creation

Editor, Arthur Fogartie | Cover Design, Joy Balogun | Children's Worksheets, Joy Lindsay

Soil & Souls Press | Marlboro, New York

Paperback: ISBN 978-0-578-209760

E-Book: ISBN 978-1-234-567897

DEDICATION

This book is dedicated to my eldest child, my daughter Arminee', who stands up for me and with me, consistently, like the sun who rises and sets faithfully every day.

Thanks to Shakeila who asked me to begin, to the listening circles, friends and family who kept me going and to Marcy who helped me finish!

CONTENTS

The Light of Christ leads me

Grace and Mercy follow me

And for that I am grateful

FOREWORD

Truly, truly, I say unto you, unless a grain of wheat falls into the earth and dies, it remains alone. But if it does, it bears much fruit. – John 12:24

Just how many of the 21st century Christians can relate to the agrarian imagery used here and other such passages in the Bible? So, do we misread scripture through our western, technologically advanced way of seeing the world? Even those that live in rural areas are now less connected to the traditional methods of farming as our food system has evolved to industrialized farming. With over half of the world's population now living in cities, such agrarian language seems less accessible and meaningful to many of today's readers of the holy scriptures.

Rev. Dele's Bible studies, reconnect today's people of faith with the regenerative agricultural practices, mindset and language needed to more deeply understand the words of scripture. She grounds her reader in the beauty of the earth and our divine purpose to care for it as it cares for us. She is breathing new life into our understanding of the Word, God, the Holy Spirit and ourselves as spiritual beings created from breath and adamah.

Written By: Stephanie Clintonia Boddie, PhD.
Baylor University, Assistant professor of Church and Community Ministries

PREFACE

We will save Earth when we love her, and we can only love her when we know her. This Bible study will introduce to you the interconnectedness of air with the other three elements and explain why that unity is important. You will also discover how to enhance your child's breath of life for generations to come. You will be challenged to connect the ground you walk on to the air you breathe to your relationship with God. You will realize that justice requires you to intertwine personal and communal balance with economic equity if our planet is to sustain human life.

The air we breathe is refreshed by the jet stream, which forms a "line dance" around the planet. I call it a "dance" because of the consistent and rhythmic ways it encircles our Earth. You may want to think of this Bible study as a dance between the Word of God and your actions in the world. When we stop paying attention, our missteps in the dance of life can cause significant harm to our children's bodies and the planet at large. Hopefully, this reading will improve your environmental dance moves. I want you to be so deeply touched that you willingly change your lifestyle in personal and public ways. May you "ease on down the road" to lead your children, homeschool, or congregation in redeeming the "breath of life" given to you and your family.

STEWARDSHIP IS LOVE

While stewardship is the process we use to care for the gifts God gives us, Creation is more than a gift—she is literally part of us. Her soil colors our bodies, her air animates our souls, and we are intended to love all of her manifestations. God created a diverse earth that responds to LOVE with built-in seasons of growth, rest, and regeneration. In the beginning, we were asked to love Earth as God loves us. In serving Earth, we preserve the foundation of abundance that nurtures both personal inner peace and communal harmony.

Because we are infused with Creation, we must not steward it as an object; we must LOVE Creation as ourselves. Right now, both Mother Earth and women are in crisis – situations that we must change. Self-care grows as a concern in our communities partially because we have neglected the invitation to grow, rest, and regenerate within a practice of love. Somehow, we have become convinced it is possible to advocate for others without caring for ourselves The result? Burnout! This study offers another approach to Creation Care.

Christ, as Lord of Creation, gathered his disciples, and reminded them to love one another as he loved them. Likewise, when this was written, sisters gathered to talk about what love and Creation Care mean to each of us.

We pray you will also meet and be inspired to grow, rest, and regenerate within a practice of love. May you be made whole within yourselves and within your communities.

Rev Dele

INTRODUCTION

Are you reeling in fear of catastrophic collapse or barricading yourself against all of "them" who are bound to get you? Breath of Life is a Bible study that invites you to shift your focus to love Earth so we can heal the inner and external conditions that make our atmosphere toxic. This 5-week study gives you assurance as well as tools to move All of us through chaos–in safety–with our neighbors.

After 40 years of church leadership inside the four walls, I spent 6 years crossing 3 continents listening to young adults yearn for a spiritual home they can trust. I learned that churches become "safe" when our ministries nurture Earth's well being and contemplative forms of prayer and practice become normal. Then your next generation of leadership will flow in like a stream.

Whether you are thirsting for fulfillment after your environmental work or merely seeking practical information for Creation Care ministry,

Breath of Life moves you closer to God through a deeper understanding of God's creation.

Veronica Kyle, outreach director for Faith in Place, says Breath of Life, "*makes me want to run and sit under a tree by a stream of water and bask in its revelations. ...I need this to continue to be inspired and encouraged to stay on this journey.*"

This Bible study will mature your faith and exponentially increase your faith community. Leave the treadmill of programs that fewer and fewer people attend and seed a new reverence for Creation. A revived you and a vibrant community begins NOW!

LEADER'S NOTES

Each chapter has a theme and focuses scripture(s). If you have a Green Bible, you will find many study aids for your class. People rarely arrive at Bible study at the same time. The Gathering Activity is given as a suggestion to discipline people's focus on spiritual matters as they wait for others to arrive. If your group prefers to do a circle "check-in" to allow everyone to share something significant in their day, that works as a fine substitute.

The opening prayer and song give everyone the opportunity to "tune in" to one another before you begin teaching. The original prayers may double as meditations. Consider singing along with the YouTube recordings. Lyrics and additional songs are included in the Reference Section. The reflection following each lesson is intended to provide personal, guided practice in the class and independent practice at home. Additional suggested activities are included for you and your child(ren), for your homeschool or playgroup, or for your congregation of faith. We hope you enjoy the corresponding children's worksheets as well.

CHAPTER 1.

CREATION

A Gospel of the Garden

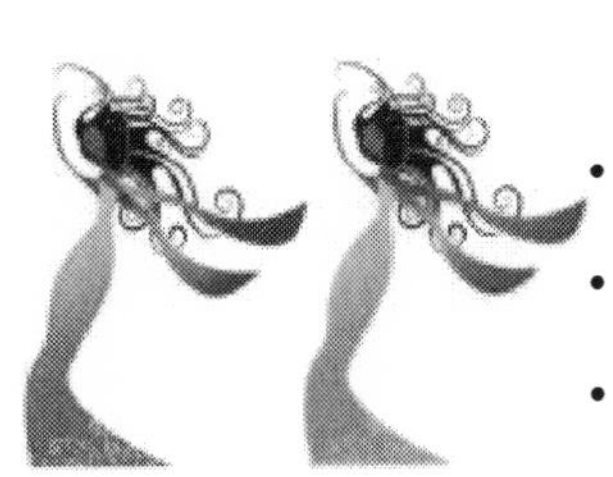

Focus Scriptures:

- Genesis 1:26-28
- Genesis 2:8-9
- Genesis 2:15

GATHERING ACTIVITY

In Genesis 2:15, what does God mean when He asks us to "serve and preserve" the Garden? God created ecosystems as "outdoor rooms" for our planet (forests, prairies, lakes, rivers, floodplains, wetlands, meadows, tundra, etc.) Spend some time guessing which outdoor room "hosts" your housing development. Does your lawn match God's original design for that area? What do you think made the air fresh in the original design? In what small ways can you help replenish the air?

Take a little time to watch and discuss the *Nature is Speaking* video. (*The video links are provided under Resources in the back of the book.*)

OPENING PRAYER

Holy, Gracious, and Loving God,

We are in awe of this beautiful planet you asked us to care for.

We wish everyone were blessed to enjoy the scent of blossoms in the air and the sense of fresh dew on the grass.

We need your wisdom to restore battered ecosystems.

We need your discipline to transform our lifestyles,

And we need your courage in crafting new policies so that everyone can freely breathe your breath of life.

SONGS

- *Breathe Into Me* by Fred Hammond
- *You are God Alone* by William McDowell

TEACHING

A Gospel of the Garden

"When God coupled Earth with the breath of eternity, our souls and the soil were fused, and our destinies perpetually intertwined." (Gospel of the Garden Sojo.net 4-23-2014) While many of us have been taught that human beings have dominion over the earth, we fail to understand that what we do to Mother Earth, we do to one another and to God. The "dominion theology" expressed in Genesis 1:26-28 has led to abuse and the destruction of Mother Earth and human communities. We have forgotten we must preserve our moral integrity to maintain the cleanliness of the air and fertility of the land. (Genesis 3:17) Bio-diversity provides a template for cultural diversity when we stop to listen and learn the lessons of the Garden.

Every time we strip the land of its diversity, we remove a layer of humanity from our collective souls. Conversely, the seven-layer forest system described in Genesis 2:8-9 represents an integrated design where soil, plants, and water work together in natural harmony to preserve clean air. In addition to the food they provide human beings and wildlife, trees increase and preserve the soil; they provide oxygen through photosynthesis and generate rainfall through transpiration. One acre of new forest can sequester about 2.5 tons of carbon annually within a soil community of beings. A single, mature chestnut tree provides more food than an acre of corn. I suggest we redeem ourselves with a "Gospel of the Garden" and recall the plant diversity God designed. Just as Christ restores humankind to union with God, let us return the land to its diversity, water to its purity, and air to its life-giving force.

Christ regularly crept away to a garden at night to nourish himself in Creation—sometimes to the Garden on the Mount of Olives (Luke 21:37, 22:39)—sometimes to the Garden of Gethsemane (Matthew 26:36, Mark 14:32. He found clarity about his life's mission, guidance for the journey, and strength to overcome obstacles. He was transfigured in the Garden (Matthew 17:1, Mark 9:2, Luke 9:28)—unified with our Creator and fortified by the prophets who had gone before. Let us follow Christ into the Garden and be fortified to restore our home—Earth. In the Garden, let us be anointed with courage and strength to overcome the forces that threaten to destroy our communities and our health.

The Garden blueprint in Genesis 2:15-16 describes a beautiful human habitat. God wants us to preserve such a stunning place. It provides shelter, food, and medicine in generous measure. God created a clean air/abundant food production system with trees as its foundation (instead of removing trees as we do). God did not segregate beautiful trees from food-producing trees, nor were crops separated from the trees. Fertilization and pollination occurred naturally. The diversity of native plants kept the number of insects in balance. The only requirements for bounty were pruning, harvesting an abundant yield, and restoring fertility to the soil.

The Hebrew words used in God's instruction are *la'avad* and *l'ishmor*. *La'avad* is the same word Joshua uses when he says "as for me and my house, we will *serve* the Lord." That means there is an attitude of reverence and joy we are to cultivate in our relationship with Earth. *L'ishmor* means to "preserve" or "keep." We have been given a system of naturally recurring abundance. Maintaining that principle of bounty represents a bedrock of our faith and our health.

REFLECTION

Reading about nature is fine, but if a person walks in the woods and listens carefully, he can learn more than what is in books, for they speak with the voice of God.
— George Washington Carver

ACTIONS

For Yourself:

Spend 10 minutes outdoors under a tree, breathing deeply and exhaling through your nose. Every time you exhale through your nose, the body naturally relaxes. Work toward rhythmic breathing where you inhale for the same amount of time as you exhale. Record how your mind shifts. The longer you sit and breathe, what can you see/hear more clearly? What challenges keep surfacing that need to be dispelled? See if you feel your shoulders, head, and body relaxing. Repeat the exercise three times this week and share the results with your Bible Study Group

With Your Child:

Teach your children to breathe rhythmically and ask them to practice when they feel particularly annoyed.

With Your Congregation:

As part of your next church meeting, go outside and spend at least five minutes among some trees and bushes. #Touchtheground and record what happens with your social network.

CHAPTER ONE:
GOSPEL GARDEN
Word Bank:
Serve
Preserve
Eco System
Mother Earth
Bio diversity
Transportation
Habit
Fun Fact
Did you know...
We globally now require the equivalent of 1.4 planets to support our lifestyles?

MY BIBLE STUDY NOTES

CHAPTER 2.

OUR INTERDEPENDENCE

Mother Wisdom Joins the Beloved Community

Focus Scriptures:

- *Proverbs 8:22-31*
- *Romans 8:19-22*

GATHERING ACTIVITY:

In this gathering activity, you will watch one of the Prince EA's video clips listed below and discuss. (*The video links are provided under Resources in the back of the book.*)

- Dear Future Generations, Sorry
- Man vs. Earth
- Will This Be Humanity's Fate

OPENING PRAYER

God of Creation, of cosmic networking,

We ask that you continue to weave wisdom into our personal and national choices.

May we cultivate Nature's power with respect, so that she will cultivate our humility.

May her beauty nurture our joy and her stillness promote our peace.

May Nature bring forth fruit of our Spirit, even as we harvest from her.

SONGS

- *How Deeply I Need You* by Shekinah Glory
- *I Need You to Survive* by Hezekiah Walker

TEACHING

Martin Luther King offered us a vision of "The Beloved Community" as an end goal of social justice work. In the Green Bible, Ellen Davis echoes that thought, "Because we have no life apart from the health of soil, water, [air], we must care for it as one would care for a beloved family member." Consider that Proverbs 8 describes what may be called the first beloved community between God, Creation, and Humanity.

Hokmah, or Lady Wisdom, is the thread that weaves God, Humanity, and Creation together; she is the relationship transformer as well. She was there in the beginning before the ecosystems of mountains, seas, and fields were formed. (8:25-26) Wisdom worked beside God "like a master worker." She rejoiced before God, rejoiced in Creation, and delighted in the human race. Notice that no one element is given preference over another because all elements must work together for each to function properly. Biodiversity is one of those wisdom threads to study as we contemplate how to be good stewards of our air.

Given that biodiversity is also a template for human cultural diversity, let's look deeper at the wisdom of interdependence within and between ecosystems. Ecosystems include family relationships among specific birds that feed on specific insects that, in turn, feed on plants native to specific soils and climates. These ancient relationships preserve habitats that maintain clean air and are "fueled" by the proximity of thousand-year-old aquifers. What happens to the interconnectedness of life when human beings use up a millennium's worth of water in only a century?

Another interdependent ecosystem preserving clean air are the rainforests. They were originally placed on every continent and function as the lungs of the planet. By anchoring soil, facilitating rainfall and filtering air particles through their leaves, rainforests represent the first line of defense for clean air. Are we exhibiting wisdom when we clear-cut 1000-year-old trees and replace a bio-diverse ecosystem with a human-driven barrage of concrete, asphalt, air pollution from vehicles, chemical plants, and coal-burning factories? How can we practically restore the presence of wisdom?

Teachers and scientists alike have noticed the connection between clean air and our mental and emotional functions. "I notice that it is only when my mother is working in her flowers that she is radiant, almost to the point of being invisible except as Creator: hand and eye." She is involved in work her soul must have. Ordering the universe in the image of her personal conception of beauty." (Alice Walker, *In Search of Our Mother's Gardens*) What does walking or working in a beautiful space do for your appreciation of God and the many interconnections we enjoy? Richard Louv also discussed the need for a "vitamin N" because Nature not only gives us air, water, and food, but also a balanced nervous system and emotional well-being.

In addition to the individual human connectedness with specific species for food or medicine, Bell Hooks invites us to appreciate the role Nature has played in community wholeness and our social justice movements. In her book *Earthbound*, she recounts, "Growing up in a world where my grandparents did not hold regular jobs but made their living digging and selling fishing worms, growing food, raising chickens, I was ever mindful of an alternative to the capitalist system that destroyed nature's abundance. In that world, I learned experientially the concept of "interbeing." Do you see evidence that our economic system serves and preserves Nature's abundance? How would you describe your concept of interbeing with Nature?

Wisdom seems to be removed from the beloved communities when mountaintops are cleared for coal mining. Air quality is quickly diminished for all plants, animals, birds, and human beings who live close by. Coal miners work for low wages and have inadequate health care (because of their higher incidence of lung disease and various cancers). When we dismantle ecosystems, their other inequities abound–a dismantling of a fair work ethic, a demise in physical health, and a disconnect in the original joy between God, humanity, and

Creation. What is the Psalmist suggesting about the relationship between air and earth: "The heavens are the Lord's Heavens, but the Earth he has given to human beings" (Psalms 115:16)? If our earthly actions create a negative impact on the heavens, have we overstepped our boundaries? No wonder the author in Romans 8:19-23 speaks of Creation's groaning with all of humanity to birth a new reality—one free from the slavery of corruption and human selfishness. What can we do to push back on the 140 million tons of coal ash, arsenic, lead, and selenium ejected into the air for all of us to breathe?

REFLECTION

Smothering smog occurs after we have inundated biodiverse habitats with monoculture lawns and other single species. When the British colonized American people, they also colonized our landscapes, planting turf that grows naturally in England. Do you think lawns "serve and preserve" the abundant yields that we were originally given? Contemplate ways we can use our landscapes to breathe more freely the breath of life.

ACTIONS

For Yourself:

Inhale deeply and exhale the "Ah" sound until you feel the resonance in your heart. Repeat for one minute. Sit quietly and reflect on how you feel before moving back into your day.

With Your Child:

Walk in the words and make a game of this slogan from the Chipko Movement in Adwani: What do the forests bear? Soil, water, and pure air! Research their work in the Himalayas.

With Your Congregation:

"How many planets do we need?" Complete the following activities at and compare notes:

- See the Footprint Calculator at www.footprintcalculator.org.

Research the Greenbelt Movement. Why do I say that Wangari Maathai is a champion for the lungs of the earth?

CHAPTER TWO:
OUR INTERDEPENDENCE
Word Bank:
Community
Social Justice
Humanity
Interbeing
Cultural Diversity
Fun Facts:
Did you know...
A giraffe and the oxpecker
have an interesting
interdependent relationship.
Every day a giraffe deals with
unwanted parasites on
a daily basis, like ticks
and flies. The oxpecker,
eats these and other
intruders off of the giraffe
Both animals help one
while giraffe gets
cleaned, the
oxpecker gets fed.

MY BIBLE STUDY NOTES

CHAPTER 3.

OUR BREATH OF LIFE

Polluted Air Anywhere, Means Polluted Air Everywhere

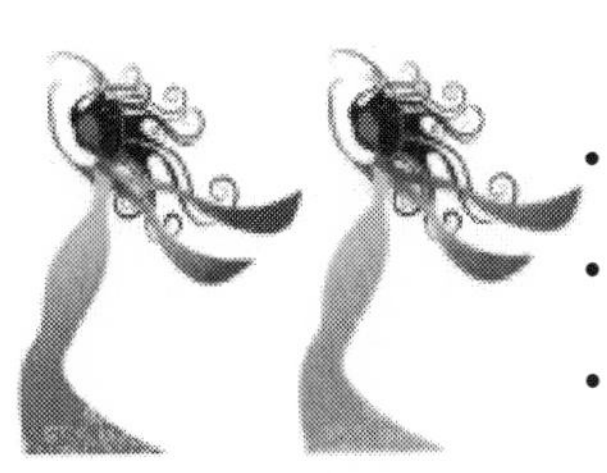

Focus Scriptures:

- Exodus 9:8-10
- Ecclesiastes 12:7
- Genesis 2:7

GATHERING ACTIVITY:

In this gathering activity, you will watch Nature is Speaking: Joan Chen is Sky video clip and discuss. (*The video link is provided under Resources in the back of the book.*)

OPENING PRAYER

Holy Spirit, Holy Spirit

Caress us with your presence

Fill our hearts with joy

Create a sacred space of love

Be With Us

SONGS

- *This is the Air I Breathe* by Michael Smith. (Close your eyes and breathe into your heart center. Observe what you see and feel. See if you can maintain sacred space even while reading the science material below.)

TEACHING

The Creation narrative in Genesis tells us that when the waters separated, the sky appeared, holding the largest amount of uninterrupted air. Air is a mixture of gases that encircle the earth (about 78% nitrogen, 21% oxygen, 0.9% argon, 0.04% carbon dioxide, and very small amounts of other gases). The custom-designed ratios are perfect for life on this planet and must stay in balance for healthy breathing to occur. Trees are the main balancing agent of human clean air.

God drew a canopy around Earth to keep that mixture close by for our use (Genesis 1:6-7) and to allow room for constant purification through the jet stream as it undulates all around the globe. After allowing air to form and making a special place for a lot of it to hang out (in the sky), God created land, vegetation, and animals until the end of the sixth day. Then, God formed human beings from dust and animated them through an intimate and special gift of this air called *nishmatchayyim* or "the breath of life" (Genesis 2:7). If you were grateful for every breath of life you experience in a day, how many times would you say, "Thank You?" How do you think your life would change if you were that grateful every day?

Since our breath is such a precious gift and such special conditions were put in place to keep it healthy, all of us should know some basic facts about air. Even after we learn the facts, how do we keep a sense of the sacred when we perform air advocacy?

One thing we know about healthy air is that it stays in constant motion while interacting with bodies of water and with leaves in forests. Being quite the "party person," air requires circulation. Large bodies of water moving with intense force help to transfer negative ions to air and make it cleaner and lighter.

Waterfalls and ocean shores are perfect places to experience the physical, emotional, and spiritual impact of freshly cleaned air. And yet, the water must itself be clean in order to reap this benefit. Warm, nutrient-poor water helps to transform wind patterns into destructive storms known as *El Nino* and *La Nina*.

Remember our interdependence in Chapter Two? In the beginning, each continent had a rainforest, and this network of forests formed the "lungs of the earth." With most of our rainforests in the United States clear cut, we find the absence of leaves makes for poorer quality of air in most of our neighborhoods. Spend ten minutes writing a statement to convince your neighbors of the importance of trees in reducing the incidence of asthma.

Maintaining carbon in the earth is a way to support clean air. Whenever the natural carbon cycle is reversed, and earth particles stay in the sky, human beings get sick and other animals die. Read what happened in Exodus 9:8-10 when soot from large furnaces was thrown into the air. While in Biblical times people developed boils, today, people develop asthma and cancers. The specific diseases may be different, but it is clear that we are all healthier when coal stays in the ground.

Air is experienced in the human body by the process of breathing. Oxygen enters the body and carbon dioxide leaves the body. This change in the composition of air is what creates the interdependence between the animal and plant kingdoms. Our lung capacity and skin work together to keep our bodies animated and toxin-free. Our capacity to hold this breath of life dictates our quality of life as well as our life span (Ecclesiastes 12:7). When the air we breathe contains more poisons than the body can eliminate, not only do our bodies become sick, but Earth becomes sick as well. Enlightened self-interest tells us we must make necessary changes to keep all of God's Creation in good working order. Discuss which act of repentance (which change) you will make this day?

REFLECTION

When Marvin Gaye wrote *Mercy, Mercy Me,* it was seen as prophetic. Have your personal actions since the 1970's made things better or worse? Are we like the people of Nineveh who repented and were spared, or like the Israelites in the Wilderness who resisted the will of God and were destroyed?

ACTIONS

For Yourself:

Use this breath to clear your energy centers from anger and frustration. Inhale deeply and exhale the sound of "Hauh." Relax, drop your jaw, and really have fun with this. See if you can feel your breath resonating in your solar plexus. It sounds a little like Darth Vader when done correctly. Repeat for 30 seconds and observe how you feel.

With Your Child:

Intensify your breathing practice to increase your awareness. Attempt to exhale through your nose so softly that you do not feel breath moving across your nostrils. Keep jaw, shoulders, and toes soft during this process as well.

With Your Congregation:

"Mother Nature must be seated at the table of the Beloved Community, not as a table decoration, but as a voice to be heard and as principles to be followed." In your church and social justice work, how can you include Nature's priorities as your priorities, and Nature's principles as your governing principles?

CHAPTER THREE:
OUR BREATH OF LIFE
Fun Facts:
Did you know...
The lungs are the only organ in the body to float in water.
Nitrogen
Carbon Dioxide
Argon
Natural carbon
Lung capacity
Repentance
Oxygen
Process of breathing
Word Bank

MY BIBLE STUDY NOTES

CHAPTER 4.

MOTHERHOOD IS CLIMATE JUSTICE

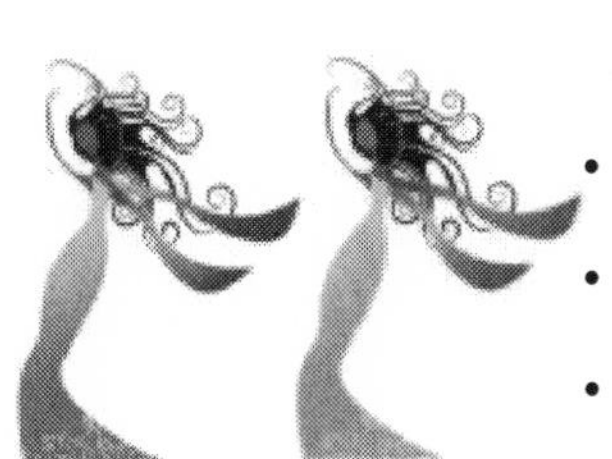

Focus Scriptures:

- Proverbs 8
- Exodus 2:2-3
- John 19: 26-27

GATHERING ACTIVITY:

Take turns reading the lines of the Universal Declaration of the Rights of Mother Earth found at www.therightsofnature.org/universal-declaration and discuss. While this document was initially prepared by nations from the global south, it is now used in the United States for environmental justice work.

OPENING PRAYER

God who is our mother and father, provider and protector,

You who gather us under your wings as a mother hen.

Continue to love us as we perform the difficult task of repentance.

Give us the humility to return as naughty children and resolve to do better,

And then grant us the strength to follow through

Amen

SONGS

- *Breathe* by Indie Arie

TEACHING

This chapter is deeply reflective and challenges you to mature as a student. Before listening to my historical research and philosophical perspectives, I want you to plumb the depths of your own intuition and spirit and come to class with a prepared response to the two passages and the questions that follow. A prepared response usually takes less time to share and has more "meat" than a totally impromptu response.

Passage One is an example of a mother's heart-rending sacrifice for her own child. Passage Two is an example of a Hebrew son's responsibility to his mother. Because the passages can become so deeply personal, you may want to employ a practice from my MicMaq ancestors that are shared among many Native Americans. In the talking circle, only the person holding the "talking stick" may speak. They speak from the heart and are not interrupted. The stick returns to the center between speakers. There are no critiques, challenges, or arguments. Sometimes, a feather is attached to the stick to symbolize the desire that one's heart and speech will be as light as a feather.

PASSAGE ONE: Exodus 2:2-3 Pharaoh/the king of Egypt has decreed that all male, Hebrew children shall be killed at birth. While the midwives did not cooperate with this order, soldiers did. Hebrew parents had to be very creative in the ways they prevented their sons' deaths. Moses' mother tried everything. She hid her newborn under blankets in corners and baskets for months, but the child soon became too large to hide. She gave him up, even though it must have broken her heart. It was better to face an uncertain freedom among alligators and snakes in the river than certain death in the house.

We hold many addictions close to us as dearly as we do our own children. The addictions that endanger the air include individual ownership of big cars, clear-cutting forests of all types, maintaining monoculture on our lawns, and monoculture in all of our crops. How can you cultivate the strength to let go of these addictions, so your community can thrive? What resistance do you face? How might working with other groups help dissolve your resistance to maintaining a healthy planet?

PASSAGE TWO: John 19:26-27 In the Hebrew patriarchal society, the eldest son became legally responsible for the safety and wellbeing of his mother once the father was gone. Having a son was critical because only men could transact the business of daily living. Realizing he is no longer able to carry out his responsibility, Jesus designates someone. Notice, he first notifies his mother of the change, "Here is your son"(verse 26) and then notifies the new son, "Here is your mother" (verse 27). There is the promise from Jesus, but the follow through comes from the new son: "From that hour, the disciple took her into his own household." Even as he feels death approaching, Jesus makes sure his mother is cared for. Surely, he could have solely focused on his extreme pain. He had been hanging in the hot sun for hours; he had been beaten all night. He could have zeroed in on the injustice of being there in the first place. Instead, Jesus ensures his mother and the generations to come that they will be cared for.

Have you ever delegated a really important responsibility? What did you consider in the process? Do you think God made those same deliberations when God delegated human beings to take care of the planet? What might God be saying now? Recall the last time you were in excruciating pain. Did you focus on just your own condition or were you able to take care of others at the same time? Can Jesus' example of caring for his mother inform the way we put aside our own comfort and care for the security of future generations?

In summary, many of the behaviors we associate with mothers represent principles for climate justice. Mothers set limits so the whole family can have enough and interact in a way that is healthy. Just as rivers flow within banks and oceans within shores, personal limits are essential for all healthy interactions either between children or between human beings and our planet. Sabbath is just another word for "time out." Mothers use time out when children have pushed past healthy behaviors and need to rebalance. Sabbath was initiated as a form of economic and ecological "time out" because ancient Hebrews were in the practice of overworking everyone and not taking time to renew either the land or one another. Sabbath was initiated to ensure justice for humans and the land. Is overwork and lack of renewal time still a problem? Why or Why not?

REFLECTION

Do you practice Sabbath now? Given the above justice definition, might you consider another way to practice Sabbath?

ACTIONS

For Yourself:

This breathing exercise focuses on clearing your throat area, so you can speak your Truth clearly and with courage. Inhale deeply and exhale the "ooo" sound (as in "soon"). Play with different tones, speeds, and pitches until you feel a sense of peace with your voice. Rest a minute before returning to your day.

With Your Child:

Discuss fun ways to keep the Sabbath, so the family can practice a culture of growth, rest, and regeneration.

With Your Congregation:

When Nature is not part of the justice culture, imbalance seeps into our social actions. We always look for speakers who are *"on fire"* and then wonder why we are burning ourselves out. How can we make program decisions that balance fiery addictions with presentations that reflect the cool and calmness of the earth and the refreshment of water?

CHAPTER FOUR:

MOTHERHOOD IS CLIMATE JUSTICE

Fun Facts:

Did you Know...

Americans use about 380 billion plastic bags each year, which balances to about 1,200 bags per person.

Word Bank:

Intuition

Sacrifice

Justice

Economic Justice

Ecological Justice

MY BIBLE STUDY NOTES

CHAPTER 5.

GOD IS CALLING

How Will You Live Out the First Commandment to Serve & Preserve Creation?

Focus Scriptures:

- Colossians 1: 16,20
- John 1:3; Hebrews 1:3
- Luke 6:46

GATHERING ACTIVITY:

Share stories about a time when you felt God calling you to do something. Did you laugh at the possibility like Sarah, avoid the call and end up with more trouble like Jonah, give God conditions like Moses, or respond eagerly like Mary?

OPENING PRAYER

Gracious One, you see through my eyes and breathe through my lungs.

Allow me to breathe freely the breath of life and to help all people to do the same.

Strengthen my inner being and give me courage to live a life of compassion within my family, strength within my community, and to perform justice that balances the world.

SONGS

- *Put a Praise on It* by Tasha Cobbs

TEACHING

In the beginning, humankind was in direct communion with God. God and human beings spoke freely and understood one another clearly (Genesis 2:16, 2:23, 3:9-19). Mother Nature provided all material needs through the Garden (Gen 2:29). The only requirement was that men and women should serve and preserve Earth. In other words, make sure Earth has what she needs to remain fruitful: Love Her. Keep your head and heart working together. If this appears difficult, ask Wisdom! Long before Moses went to Mount Sinai, this was the first commandment.

One act of consumption brought a curse to Earth and her inhabitants (Gen 3:17). But praise be to God, 2000 years ago, Christ, as Lord of Creation, redeemed us from this curse. No longer must any aspect of Creation suffer from domination—not people, soil, waters, or air. In a sense, we are back to the beginning, and all we have to do is convince and redeem whomever "didn't get the memo." The "memo" said God's way of living on the land is best. If we want fresh air and abundance to return to the land, we must redeem bio-diversity—in our front and back yards, public right of ways, and croplands. Governments must also be held accountable for the parts of life they hold. That means we must speak to our politicians about destroying ecosystems with pipelines, polluting the air with coal, and installing tree canopies to bring breath back into our neighborhoods.

Those of us who live, breathe, and have our being in Christ Jesus (Acts 17:28) also live, breath and have our being in Creation. In order for that "being" to be healthy, we must have clean air. Disciples of Christ cannot ignore Creation, because "all things have been created through him and for him" (Colossians 1:15-16). All of Creation is held together by Christ's consciousness. Since Christ is "Lord of Creation," respecting Creation is the same as respecting Christ. Discipleship—following Christ—includes protecting Creation. When we plant trees, we are preserving the air and following Christ. When we petition our leaders to clean up air pollution, we are acting as faithful stewards in redeeming Earth. When we model a work ethic that has breathing room, we are observing God's Sabbath. Will you answer Wisdom's call to restore our earth or will your inaction bring forth this rebuke, "Why do you call me Lord, Lord and do not do what I tell you?" (Luke 6:46)

REFLECTION

Finish sharing your past call stories. Discuss what God is calling you to do now?

ACTIONS

For Yourself:

This breath brings soothing calmness to your mind. It is the simple "m" sound you hear when humming. Remember our fore parents used to hum as they were working? Inhale deeply and exhale the "m" sound as you smile. Repeat for a minute or as long as you like. Rest and reflect before resuming daily activity.

With Your Child:

Ask your child what their special talents are right now and how those skills may be used for the glory of God. Ask if they ever hear that "still small voice" and how it differs from other voices.

With Your Congregation:

Caring for Creation is a blueprint for community care. Brainstorm on ways you can demonstrate this truth in your community. How will you include spiritual reflection while you are physically caring for your garden?

CHAPTER 5:

GOD IS CALLING

Fun Facts:

Did you know...

A vegetable garden is a man-made ecosystem. Plants and flowers entice plant eating insects. These insects in return attract birds, snakes , frogs and other predatory insects. These animals might attract even bigger predators such as foxes, racoons, coyotes, and birds of prey.

MY BIBLE STUDY NOTES

RESOURCES

LINKS FOR GATHERING ACTIVITIES

Chapter 1:

- Nature is Speaking — https:/youtu.be/WmVLcj-XKnM

Chapter 2:

- Dear Future Generations, Sorry — https://youtu.be/eRLJscAlk1M
- Man vs. Earth — https://youtu.be/VrzbRZn5Ed4
- Will This Be Humanity's Fate — https://youtu.be/mRUoLarLLVA

Chapter 3:

- Nature is Speaking: Joan Chen is Sky — https://youtu.be/E8d_JvMpoY4

Chapter 4:

- Universal Declaration of the Rights of Mother Earth — http://www.therightsofnature.org/universal-declaration/

ECOLOGY QUOTATIONS

Quotations by Rev. Martin Luther King, Jr.

1. "It really boils down to this: that all life is interrelated. We are all caught in an inescapable network of mutuality, tied into a single garment of destiny. Whatever affects one destiny, affects all indirectly."
2. "Never, never be afraid to do what's right, especially if the well-being of a person or animal is at stake. Society's punishments are small compared to the wounds we inflict on our soul when we look the other way."
3. "Our scientific power has outrun our spiritual power. We have guided missiles and misguided men."
4. "We must rapidly begin the shift from a 'thing-oriented' society to a 'person-oriented' society. When machines and computers, profit motives and property rights are considered more important than people, the giant triplets of racism, materialism, and militarism are incapable of being conquered."
5. "I just want to do God's will. And he's

allowed me to go to the mountain. And I've looked over, and I've seen the promised land! I may not get there with you, but I want you to know tonight that we as a people will get to the promised land."

6. "From the prodigious hilltops of New Hampshire, let freedom ring. From the mighty mountains of New York, let freedom ring. From the heightening Alleghenies of Pennsylvania, let freedom ring. But not only that: Let freedom ring from every hill and molehill of Mississippi."
7. "Only in the darkness can you see the stars."
8. "We have flown the air like birds and swum the sea like fishes but have yet to learn the simple act of walking the earth like brothers."

Quotations by Alice Walker

This is from In Search of Our Mothers' Gardens: Womanist Prose. San Diego: Harcourt Brace Jovanovich, 1983.

"In Search of my mother's garden, I found my own."

"I notice that it is only when my mother is working in her flowers that she is radiant, almost to the point of being invisible except as Creator: hand and eye. She is involved in work her soul must have. Ordering the universe in the image of her personal conception of beauty."

Quotations by Bell Hooks

This is from Chapter 10 of the book Belonging: A Culture of Place, Taylor and Frances, 2008 (Page 67).

"A child of the hills, I was taught early on in my life the power in nature. I was taught by farmers that wilderness land, the untamed environment, can give life and it can take life. In my girlhood I learned to watch for snakes, wildcats roaming, plants that irritate and poison. I know instinctively; I know because I am told by all-knowing grown-ups that it is humankind and not nature that is the stranger on these grounds. Humanity in relationship to nature's power made survival possible. "

"In that world, country black folks understood that though powerful white folks could dominate and control people of color, they could not control nature or divine spirit. The fundamental understanding that white folks were not gods (for if they were they would shape nature) helped imbue black folks with an oppositional sensibility. When black people migrated to urban cities, this humanizing connection with nature was severed; racism and white supremacy came to be seen as all-powerful, the ultimate factors informing our fate. When this thinking was coupled with a breakdown in religiosity, a refusal to recognize the sacred in everyday life, it served the interests of [the] white supremacist capitalist patriarchy."

"Living in the agrarian south, working on the land, growing food, I learned survival skills similar to those hippies who sought to gain their back-to-the-earth movements in the late sixties and early seventies. Growing up in a world where my grandparents did not hold regular jobs but made their living digging and selling fishing worms, growing food, raising chickens, I was ever mindful of an alternative to the capitalist system that destroyed nature's abundance. In that world, I learned experientially the concept of interbeing, which Buddhist monk Thich N'hat Hanh talks about as that recognition of the connectedness of all human life."

"That sense of interbeing was once intimately understood by black folks in the agrarian south. Nowadays it is only those of us who maintain our bonds to the land, to nature, who keep out vows of living in harmony with the environment, who draw spiritual strength [from] nature. Reveling in nature's bounty and beauty has been one of the ways enlightened poor people in small towns all around our nations stay in touch with their essential goodness even as forces of evil, in the form of corrupt capitalism and hedonistic consumerism, work daily to strip them of their ties with nature."

"More than ever before in our nation's history, black folks must collectively renew our relationship to the earth, to our agrarian roots. For when we are forgetful and participate in the destruction and exploitation of the dark earth, we collude with the domination of the earth's dark people, both here and globally. Reclaiming our history, our relationship to nature, to farming in America, and proclaiming the humanizing restorative of living in harmony with nature so that the earth can be our witness is meaningful resistance."

Dr. Melanie Harris: "Eco-womanism" centers the voices and perspectives of women from African descent as they engage earth justice. Drawing upon African cosmology these perspectives understand the earth as sacred. There is a connection between humans, the divine realm, and the earth. This worldview guides ethical decision-making, community relationships, and understandings of health and welfare. Showing care and love for the earth are ways in which beings can live out principles of mutuality, reciprocity, and equality in relationships; human to human, and human to non-human. "

"Uncovering one's eco-memory is the first step of an eco-womanist method. Essentially this invites you to reflect upon your own relationship with the earth and to recall a story from your childhood, or your life since in which you experienced a connection to nature."

"There is a connection between the suffering of women and the suffering of the earth. There is a connection between the logic that allows for these oppressions to take place in human and non-human (earth) community. Tracing that connection, (in this case) in the logic that purports racial superiority and gender oppression to the logic that insists the earth be controlled and dominated, is a part of the work of Eco-womanism. When we look towards solutions in the form of racial justice and gender justice we are also simultaneously looking for solutions for earth justice."

Rev. Dr. Delores Williams makes this point in an article entitled "Sin, Nature and Black Women's Bodies" wherein she documents how African-enslaved women's bodies were used for breeding. They were raped, sexually violated, and impregnated, some being forced to give birth to multiple sets of children during the course of one lifetime. (Some slave narratives record enslaved women giving birth to more than 24 children.) Williams suggests there is a striking parallel between the logic at work here and the process of strip-mining that can dangerously impact a
mountain's ability to produce or reproduce coal. Just as the sacred womb of a woman is trauma-

tized by rape, so too is the womb of a mountain traumatized by an act of earth violence.

Quotations by George Washington Carver

"I love to think of nature as an unlimited broadcasting station, through which God speaks to us every hour if we will only tune in."

"My prayers seem to be more of an attitude than anything else. I indulge in very little lip service, but ask the Great Creator silently, daily, and often many times a day, to permit me to speak to Him through the three great Kingdoms of the world which He has created—the animal, mineral, and vegetable Kingdoms—to understand their relations to each other and our relations to them and to the Great God who made all of us. I ask Him daily and often momently to give me wisdom, understanding, and bodily strength to do His will; hence I am asking and receiving all the time."

"Reading about nature is fine, but if a person walks in the woods and listens carefully, he can learn more than what is in books, for they speak with the voice of God."

"If you love it enough, anything will talk with you."

SONG LYRICS

Breathe Into Me – Fred Hammond

When the battle makes me weary
It seems that I've lost ground
It's so hard to hear Your voice Lord
With distractions all around
I try to lift my hands, to give You praise
But then a spirit of heaviness
Tries to shield Your face, so I'm saying breathe

Breathe into me oh Lord, the breath of life
So that my spirit would be whole
And my soul made right
Breathe into me oh Lord, day by day
So that my heart is pure before You, always, always

Unto Thee oh Lord do I lift up my soul
It's only by Your hand
That I can be made whole
So Lord breathe on me
And revive my spirit within
And I'll never be the same

Breathe into me oh Lord, the breath of life
So that my spirit would be whole
And my soul made right
Breathe into me oh Lord, day by day
So that my heart is pure before You, always, always

Condemnation tries to hold me
Like a prisoner in chains
And the weight of my burdens
Is calling out my name
That's when I lift up my voice worship You
Because I know
You promised You would wash me
Jesus wash me white as snow

Unto Thee oh Lord do I lift up my soul
It's only by Your hand, that I can be made whole
So Lord let Your breath revive me again
And I'll never be the same
I will never be the same, I will never be the same

Breathe into me oh Lord, the breath of life
So that my spirit would be whole
And my soul made right
Breathe into me oh Lord, day by day
So that my heart is pure before You, always, always

Songwriters: David P. Ivey / Fred Hammond

You Are God Alone – William McDowell

You are not a God, created by human hands
You are not a God, dependent on any mortal man
You are not a God, in need of anything we can give
By Your plan, that's just the way it is

You are not a God, created by human hands
You are not a God, dependent on any mortal man
You are not a God, in need of anything we can give
By Your plan, that's just the way it is

You are God alone from before time began
You were on Your throne
You are God alone and right now
In the good times and bad
You are on Your throne
And you are God alone

You're the only God whose power none can contend
You're the only God whose name and praise will never end
You're the only God who's worthy of everything we can give
You are God and that's just the way it is

You are...

Songwriters: Phillips, Craig & Dean

How Deeply I Need You — Shekinah Glory Ministry

Here is my heart, I give it lord to you
Here is my life, I lay it before you
Where else would I go?
What else would I do?
If I did not know you?
How deeply I need you
How deeply I need you, my lord
How deeply I need you, my lord
Like the desert needs the rain, I need you
Like the ocean needs the streams, I need you
Like the morning needs the sun, I need you
Lord you are my only one
In every way, and every day
I need you

Songwriters: Darrell Patton Evans

I Need You to Survive – Hezekiah Walker

I need you, you need me.
We're all a part of God's body.
Stand with me, agree with me.
We're all a part of God's body.
It is his will, that every need be supplied.
You are important to me, I need you to survive.
You are important to me, I need you to survive.
(repeat 3X)

I pray for you, You pray for me.
I love you, I need you to survive.
I won't harm you with words from my mouth.
I love you, I need you to survive.
(repeat 8 X)

It is his will, that every need be supplied.
You are important to me, I need you to survive.

Breathe – Byron Cage

This is the air I breathe
This is the air I breathe
Your holy presence living in me

This is my daily bread
This is my daily bread
Your very word spoken to me

And I I'm desperate for you
And I I'm lost without you

This is the air I breathe
This is the air I breathe
Your holy presence living in me

This is my daily bread
This is my daily bread
Your very word spoken to me

And I I'm desperate for you
And I I'm lost without you

And I I'm lost without you
And I I'm desperate for you

And I I'm lost without you
I'm lost without you
I'm lost

Songwriter: Michael W. Smith

Breathe – India Arie

Sometimes you just can't believe the things your eyes see
So much injustice in this life
If it's happenin' right on your TV screen
So you drop to your knees and you're prayin'
'Cause you can hear him sayin' he can't breathe
And it's all so overwhelming
Because you know there's nothing you can do to help him

[Chorus]
Continue to breathe
Continue to breathe
In times like these
That's what your heart is for
Continue to breathe
Continue to breathe
In honor of your brother
That's what your heart is for

[Verse 2]
There's always someone tryin' to take someone's power away
The history of the world is violent, will it ever change?
Now we're livin' in a time where you just can't hide
There's a camera in every hand
It's not elusive, even when they treat you like

you're useless
We know what the truth is

[Chorus]
Continue to breathe
Continue to breathe
In times like these
That's what your heart is for
Continue to breathe
Continue to breathe
In honor of your brother
That's what your heart is for

[Bridge]
Fight for your life
Fight for your life
In the face of a society
That doesn't value your life
For the men in your life
For the boys in your life
For your brothers, for your fathers
For the ones that came before us
For the future, for the future
For the future, for the future

[Chorus]
Continue to breathe
Continue to breathe
In times like these
That's what your heart is for
Continue to breathe

Continue to breathe
In honor of your brother
That's what your heart is for
Continue to breathe
Continue to breathe
In times like these
Nothing matters more
Continue to breathe
Continue to breathe
In honor of your brother
That's what your heart is for

Put A Praise On It – Tasha Cobbs

There's a miracle in this room
With my name on it
There's a healing in this room
And it's here for me

There's a breakthrough in this room
And it's got my name on it
So I'm gonna put a praise on it
I'm gonna put a praise on it
Somebody put a praise on it
There's a miracle in this room
With Tasha's name on it
You outta put your name in the atmosphere, c'mon

There's a healing in this room
And it's is here for me
There's a breakthrough in this room
And it's got my name on it

So I'm gonna put a praise on it
Somebody put a praise on it
Can you help me put a praise on it
Somebody put a praise on it
Let's go
(Tasha talking… I dare somebody to put a praise in this room..)

This is My Father's World – Michael Curb Congregation

This is my Father's world,
And to my listening ears
All nature sings, and round me rings
The music of the spheres.
This is my Father's world:
I rest me in the thought
Of rocks and trees, of skies and seas—
His hand the wonders wrought.
This is my Father's world:
The birds their carols raise,
The morning light, the lily white,
Declare their Maker's praise.This is my Father's world:
He shines in all that's fair;
In the rustling grass, I hear Him pass,
He speaks to me everywhere.This is my Father's world:
O let me ne'er forget
That though the wrong seems oft so strong,
God is the Ruler yet.
This is my Father's world:
Why should my heart be sad?
The Lord is King: let the heavens ring!
God reigns; let earth be glad!

How Great Thou Art – CeCe Winans

O Lord my God,
When I in awesome wonder
Consider all
The works Thy Hand hath made,

I see the stars,
I hear the mighty thunder,
Thy pow'r throughout
The universe displayed,

Then sings my soul,
My Savior God, to Thee,
How great Thou art!
How great Thou art!

Then sings my soul,
My Savior God, to Thee,
How great Thou art!
How great Thou art!

When through the woods
And forest glades I wander
I hear the birds
Sing sweetly in the trees,

When I look down
From lofty mountain grandeur
And hear the brook
And feel the gentle breeze,

Then sings my soul,
My Savior God, to Thee,
How great Thou art!
How great Thou art!

Then sings my soul,
My Savior God, to Thee,
How great Thou art!
How great Thou art!

When Christ shall come,
With shouts of acclamation,
And take me home,
What joy…

What a Wonderful World – Louis Armstrong

I see trees of green, red roses too
I see them bloom for me and you
And I think to myself what a wonderful world

I see skies of blue and clouds of white
The bright blessed day, the dark sacred night
And I think to myself what a wonderful world

The colors of the rainbow so pretty in the sky
Are also on the faces of people going by
I see friends shaking hands saying how do you do
They're really saying I love you

I hear babies crying, I watch them grow
They'll learn much more than I'll never know
And I think to myself what a wonderful world
Yes, I think to myself what a wonderful world

Thiele & Weiss -composers

Wake Up Everybody, Part One – Harold Melvin & the Blue Notes with Teddy Pendergrass

Wake up everybody no more sleepin' in bed
No more backward thinkin' time for thinkin' ahead
The world has changed so very much
From what it used to be
There is so much hatred war an' poverty
Wake up all the teachers time to teach a new way
Maybe then they'll listen to whatcha have to say
'Cause they're the ones who's coming up and the world is in their hands
When you teach the children teach 'em the very best you can

The world won't get no better if we just let it be
The world won't get no better we gotta change it yeah, just you and me

Wake up all the doctors make the ol' people well

They're the ones who suffer an' who catch all the hell
But they don't have so very long before the Judgment Day
So won'tcha make them happy before they pass away
Wake up all the builders time to build a new land
I know we can do it if we all lend a hand

The only thing we have to do is put it in our mind
Surely things will work out they do it every time

The world won't get no better if we just let it be
The world won't get no better we gotta change it yeah, just you and me

Change it yeah, change it yeah, just you and me
Change it yeah, change it yeah
Can't do it alone, need some help ya'll
Can't do it alone
Can't do it alone yeah, yeah
Wake up everybody, wake up everybody
Need a little help ya'll
Need a little help
Need some help ya'll

Change the world
What it used to be
Can't do it alone, need some help
Wake up everybody
Get up , get up, get up, get up
Wake up, come on, come on
Wake up everybody

Songwriters: Victor Carstarphen / John Whitehead / Gene McFadden

AFTERWORD

A Tag Team conversation too good not to share!

The stands are full and the bases are loaded. A hush goes over the crowd as the batter takes his final warm-up swing and firmly plants himself over home plate. All eyes are on the ball as it spins perfectly into "the zone". The bat cracks as it connects with the ball and sends it whirling past 2nd base into the outfield. Cheers turn to gasps as the crowd helplessly watches the batter sprint and slide into second base. The outfielders are laughing and team members arguing as the Umpire trots up and definitively yells "YOU'RE OUT!" The batter shakes his head as he steps sadly back to the dugout. His teammates can't believe they have forfeited at least 3 home runs.

Bases are also loaded in our global climate situation. Too often, preachers, teachers, and activists want to skip to the second base of rules and regulations in Exodus or the policies and procedures in

Deuteronomy. And what happens? We are OUT of alignment with God's abundance and as the batter can't figure out why. Like the crowd, the heavenly host watches helplessly when we use our dominion to destroy God's creation.

When we don't take the time to *commune* with the breath of God in Nature, we forfeit the moral relationship that would prevent us from fouling the air we breathe or polluting the waters we drink. Let's change our morality before we attempt to change our laws. Stop running to second base!

The birth attendant ensures we have entered earth by watching our FIRST breath. Our families watch us take our LAST BREATH before we leave Earth. The Breath of Life is critical to understanding our right relationship to God and God's creation. We live, **breathe** and have our being in Christ Jesus, so let's first honor the Breath of Life and STOP RUNNING TO SECOND BASE!

– Rev. Dr. Gerald L. Durley

ABOUT THE AUTHOR

Rev. Dele is a grandmother, author, and pastor, who blends Creation Care and contemplation to train the next generation of mission leaders in faith, ecology, and economic empowerment.

She currently consults with Baylor University on Eco-theology and has been a visiting professor in Permaculture at College of William & Mary and in Eco-theology at Virginia Union University. She serves on the UN Decade of African Diaspora-Earthcare Coalition; UCC Council for Climate Justice; council member for the National Congress of Black American Indians.

Dele has been an entrepreneurial manager in California, Oklahoma, Washington DC and Virginia, designing and launching programs with national impact for over 40 years. She launched Soil &Souls to train 300 mission leaders from 30 U.S. communities that need the most assistance in climate resilience. Mission partners are emerging in South Carolina, North Carolina, Dallas, D.C., and Atlanta. Pastor Dele also co-hosts the Spirit of Resilience radio show that shares inspiration and information for our changing social and weather climates.

Dele's B.A. is from University of California-Riverside and M.Div. from Howard University School of Divinity. She serves in the Southern Conference of the United Church of Christ and maintains Baptist affiliations.

SELECTED ARTICLES AUTHORED BY REV DELE (EXCEPT WHERE NOTED):

https://sojo.net/articles/gospel-garden "Gospel of the Garden"

http://www.ecotheo.org/2015/09/fall-2015/ page 30 "Redeeming the Soil requires Redeeming the Soul"

http://greenthechurch.org/manifesto-for-food-sovereign-churches/ "Manifesto for Food Sovereign Churches"

http://www.ucc.org/pollinator_contin-
uum_of_climate_work "The Continuum of Climate Work"

http://www.ucc.org/pollinator_min-
istry_thats_good_for_the_soil_and_good_for_the_soul
An interview

http://www.geo.coop/story/earthcare-warrior-
women. An article about her work

http://www.ucc.org/a_sustainable_race_open-
ing_new_streams_of_justice_in_our_environ-
mental_and_food_systems A conference she pro-
duced in North Carolina

NOW IT'S YOUR TURN

Self-Publishing School helped me, and now I want them to help you with this FREE WEBINAR! If you have ever thought about writing a book, you CAN write a bestseller and NOW may be your time. Self-Publishing School has tools and experience across various niches and professions, so **CHECK IT OUT**!

Watch this FREE WEBINAR now, and Say "YES" to becoming a bestseller: https://xe172.isrefer.com/go/affegwebinar/bookbrosinc6644Affiliate Link/.

Made in the USA
Middletown, DE
23 December 2018